The Private Press in Leicestershire

Ann Morris

THE PLOUGH PRESS

This book has been set (with a touch of irony) on IBM Composer in 11 pt and 10 pt Press Roman and printed by offset lithography.

ISBN 0 902813 06 4

Published by The Plough Press,
28 Pantain Road, Loughborough, Leicestershire

Designed by Lynn Ball
Produced in Great Britain by Tecmedia Ltd, Loughborough

500 copies of this book have been printed for sale. Of these 100 numbered copies have been published in a special binding.

Contents

Introduction

The term 'private press' has long evaded precise definition. John Carter in his introduction to the *English Private Press 1757–1961* wrote 'The fundamental principle of private press printing, the principle that, whether or not the press has to pay its way, the printer is more interested in making a good book than a fat profit. He prints what he likes, how he likes, not what someone else has paid him to print. If now and then he produces something more apt for looking at and handling than for the mundane purpose of reading, remember he is concerned as much with his own pleasure as with yours'. The County of Leicester is unusual in containing a relatively large number of private presses which started after the second world war. I have included all the private presses listed in *Private Press Books,* published annually since 1960, and any others which I have been able to discover. My aim was to investigate the history and objectives of the presses, and their publications up to the end of 1975. In so far as possible the information was gathered by interviews with the owners of the presses, without whose help the work could not have been completed and for whose time and trouble I am most grateful. While articles have been published, in America and Malta, as well as in this country, about Leicestershire private presses this is the first attempt to bring all the information about them together in one monograph.

Garswood Press

The growth of private press printing in Leicestershire developed mainly after the Second World War. However, before 1939 there was in existence the Garswood Press, which operated in Leicester around 1936. The press was run by Faulkner Scott and A. E. Christopherson and produced one book, *The wisdom of Andrew Boorde,* being an abridged version of the *Dyetary of Helth,* 1542. The illustrations were by Christopherson. The press produced more ephemeral pieces such as the programme for a one-act play by Faulkner Scott produced by the New Stage Club, *Crowns May Perish.*

The programme was illustrated by 'S. H.' and according to Rigby Graham, who has a small collection of Garswood productions, this could be Stanley Hickson or Sam Haile. Sam Haile, an artist and actor, was killed in a car crash. A retrospective exhibition of his work was held in Leicester Museum about eight years ago. Another possible Garswood Press item, *Aria da Capo,* 1936, is a french-folded sheet printed in green and pink, with a cover designed by Stanley Hickson. There is no imprint, making its origin debatable.

Rosemary Press

In 1947, the Rosemary Press was established by Dr Albert Outram, a teacher of English. The press was little known in the world of printing because of its sparse output. Dr Outram had been interested in printing for many years but he did not actually own a press until he was in his forties. In the autumn of 1947, he started to print on a small Adana press, using 10 pt Gloucester type. The press was named after his wife.

His aim was to produce books of an imposing size with first-rate printing. Dr Outram was dismayed by the difficulty of obtaining good textbooks for sixth formers to use as a basis for discussion and criticism, so he decided to print such a book. Realising the limitations of the Adana for book work, he bought a foolscap folio Albion, which was ninety years old, but able to print two pages at a time. The final length of his *Anthology* was 154 pages, with two engraved portraits.

The *Anthology* was never used in the class, because of changes in the curriculum, but Dr Outram used it himself.

Whilst printing the *Anthology,* Dr Outram realised how little he knew about machining and tried to enrol at Leicester College of Art to learn letterpress machining and presswork. However, he was prevented because of printing union regulations. Eventually he was admitted to the typographic and practical composing classes and Tom Westley, the lecturer in charge, gave him a job for which he was paid £2. Westley had been commissioned to produce two pages of the *Revised Prayer Book* of 1928. Three copies were printed by Dr Outram with two-colour initials.

In 1951, Doctor Outram produced, also for use in the classroom, *Exercises in punctuation and rules for punctuation.* He became interested in producing his own Christmas cards, the first appearing in 1951. In 1952 he produced a further run of cards which were sold to Clarke & Satchell, a Leicester bookseller.

After some time, Dr Outram produced a prospectus which included specimen printing and a descriptive list of titles which he planned to print. He intended to produce a quarto and octavo copy of each edition, and the order of publication would depend on public demand. His prices were incredibly modest yet response proved to be non-existent so the project was abandoned.

In 1954, he planned to produce three uniform volumes, large in format, setting out his basic opinions. They were to be entitled *The Limitations of Science, Mysticism* and *A Short History of Philosophy.* The only one to be completed was *The Limitations of Science. Mysticism* was half printed and the third remained in manuscript form.

The Limitations of Science took two years to complete. It was begun on the Albion in 1958 but the press was so old and fragile that every time a sheet was printed, it required repairing. Dr Outram therefore bought, by instalments, a demy-folio Wharfedale cylinder press. His presswork was incredible considering that his time was limited through illness. The second book in the trilogy was a rewritten version of an address to a grammar school society on *Mysticism and English poetry.* It was begun in 1960 and printing continued for the next two years. The work was never completed. Dr Outram reached page 44 of an estimated 70 pages when he contracted hemiplegiac paralysis which rendered his left arm almost completely useless. Trevor Hickman of the Brewhouse Press wanted to complete the book but Dr Outram decided against this and the project was abandoned.

The Rosemary Press was probably the most scholarly press operating in Leicestershire. Dr Outram possessed very high ideals which were successfully realised in his productions.

Twelve by Eight Press

The history of the Twelve by Eight Press is inextricably tied up with its founder, John Mason and his love of fine papermaking.

John Mason is the son of a famous printer, J. H. Mason, who in his earlier years was chief compositor at the Doves Press in Hammersmith. J. H. Mason was one of the first recipients, along with such illustrious names as Eric Gill and Douglas Cockerell, of the title 'Royal Designer for Industry'. This award in 1938 was in recognition of his work in bringing private press standards to commercial printing.

Following in his father's footsteps, after leaving school John Mason went for training in fine book production to the Central School of Arts and Crafts in Southampton Row, London.

He then served an apprenticeship with Roger de Coverley the fine binders in London's West End and then obtained positions as a binder and book restorer at Oxford and then at Eton College.

Next he went to North Wales to help found the famous Gregynog private press and here for several years worked as type caster, compositor, pressman and binder. Later he went to work as manager of the Shakespeare Head Press at Stratford-on-Avon for a short period.

John Mason's father had introduced him to the art of papermaking on several visits to Germany to help Count Kessler with his Cranach Press. Kessler wanted better paper than was available at the time and was willing to pay for that privilege. John Mason's father brought back to England exquisite sheets of paper made from many different materials, including silk. John Mason was determined that some day he too would make by hand a variety of such papers for the printing of fine books.

In 1930 he became a full-time teacher of bookbinding at Leicester College of Art. He started to experiment with hand-made paper in 1954 at his home in Leicester. He made a rough mould and deckle and produced his first paper, using monbretia leaves from the garden. This was the start of a crusade to make paper at the college. Eventually, his papermaking activities were transferred to the college and he encouraged the students to search for unusual raw materials. They learned how to make sheets by the duplex method, coloured on one side and white on the other, forming excellent endpapers. One or two papers he found had coloured cotton in them, and soon they were making 'pictures in pulp'. In 1955 Mason started to experiment with nylon and eventually succeeded in producing the first sheet of nylon paper to be made in Europe.

Papermaking provided John Mason with a great deal of pleasure and the idea of the Twelve by Eight Paper Mill emerged about this time. In 1957 John Mason gave a lecture to the Double Crown club in London. This was subsequently published as *Twelve by Eight – some adventures into papermaking.* The first edition was printed and published by Leicester College of Art in 1958 on machine-made paper and with illustrations by Rigby Graham. A different edition was published by the Private Libraries Association in 1959. This was printed on Twelve by Eight paper by the John Roberts Press.

In 1959, Faber and Faber published John Mason's *Papermaking as an artistic craft.* The Faber book contained a foreword by Dr Dard

TWELVE BY EIGHT

a paper read before
the Double Crown Club
BY JOHN MASON
and now reprinted
for members of
The Private Libraries
Association

1 9 5 9

Original on Twelve by Eight paper

Hunter and 87 line block reproductions of drawings by Rigby Graham on paper from a wide variety of sources. In the same year Maggs of London published *Some papers made by John Mason* – completely on papers from the Twelve by Eight Paper Mill.

Following the publication of this book, and the subsequent editions of it published by the Twelve by Eight Press, John Mason's paper became famous and widely sought. This book is generally regarded as Mason's best work.

The Twelve by Eight Press, established around 1960, is rather different from many of the Leicestershire private presses in that it invites other creative presses to print on its paper. The paper has been particularly well used by the Stanbrook Abbey Press. Therefore the Twelve by Eight Press essentially publishes books printed elsewhere – sometimes by commercial printers and sometimes by other private presses, such as the Orpheus Press, and the Rampant Lions Press of Will Carter in Cambridge.

One of the productions was Israel Perkins' *In favour of bundling*, 1960, an old ballad illustrated by Rigby Graham. The paper was certainly different – it was produced from old bed-sheets, pulped in a tree-trunk mortar! Douglas Martin set the type, which was locked in the chase and printed by Adams Brothers & Shardlow, Leicester.

John Mason R.D.I.: a selection from the notebooks of a scholar-printer, made by his son, John Mason followed in 1961. It was justified by hand at the Orpheus Press and printed by William Caple & Company of Leicester. The text briefly covers, in J. H. Mason's own words, his time as compositor at the Doves Press and his work in helping Count Kessler set up the Cranach Press in Weimar, Germany.

The next production of the Twelve by Eight was *The papermaker's craft.* It was freely translated into English by Oliver Bayldon and printed by Will Carter of Cambridge. It contained a translation of the seventeenth century Latin poem, *Papyrus* by Father Imberdis. *More papers hand-made by John Mason* was also published around this time. It contained specimens of John Mason's paper printed at various presses. The book was begun in 1958 but the first copies were not completed until 1965, and binding continued until its eventual release in 1968.

Twelve by Eight has produced several books by Joseph Guy Lubbock. The first, in 1967, was entitled *Art and the Spiritual Life.* This was a collection of prints on Christian themes. It was bound in Linson vellum by John Mason and George Percival and priced at 21 guineas. The other books by J. G. Lubbock were *Aspects of Art and Science* 1969, and *Reflections from the Sea* 1971. *Aspects of Art and Science* was printed on Richard de Bas white laid hand-made paper. The endpapers were hand-made by John Mason. The themes of the prints included aspects of nature which had drawn the artist's thoughts, and the accompanying notes considered the relationship between the spheres of art and science. *Reflections from the sea* was printed by Will Carter and bound by George Percival.

In 1969 John Mason joined Gordon Atkins and George Percival to form Triad Publications. It was potentially a very brilliant relationship but only one book was produced, called *Princes of Wales.* It was intended to coincide with the investiture of Prince Charles at Caernarvon. However, the book was delayed and Triad publications disbanded soon after publication.

A remarkable discovery was made by John Mason when he found a parcel containing 24 copies of *Cupid and Psyche,* privately printed by J. H. Mason in 1935. This was the book on which John Mason had first learnt to set type and he was paid sixpence a page for doing so.

Since his retirement, John Mason has produced several books. He has also continued to produce copies of his sample books, which require painstaking hand work, in response to demand from all over the world. At the time of writing no special plans have been made for future productions. But in July 1976 a link with the 1920s was re-awakened. John Mason, at the age of 75, has received a special order for end papers from the newly-revived Gregynog Press.

Orpheus Press

The Orpheus Press was set up in 1958 by Douglas Martin and aptly described itself as 'the fugitive private press'. Douglas Martin was a student in the printing school at Leicester College of Art. It was here that he met Rigby Graham, a lecturer, who later contributed to the Orpheus Press.

To begin with, the press worked without equipment or premises of its own, producing occasional pamphlets on borrowed presses, including the equipment at the art school. Douglas Martin's aim was to put his own ideas on book design into practice. The first booklet that Orpheus produced was *Papermaking: John Mason,* 1958, written and printed by Martin, with 17 copies on Twelve by Eight paper and 24 copies on Abbey Mills Greenfield paper.

The next publication was *Die Sonette an Orpheus,* 1959, by Rainer Maria Rilke. This was by far the finest book Orpheus produced; it was printed on Mason's paper and included 12 three-colour lithographs by Rigby Graham. The binding, in full legal buckram, took a considerable time, and three other pamphlets were produced meanwhile. These were John Clare's *Lines written in Northampton County Asylum,* Chidiock Tichbourne's *My Prime of Youth is but a Frost of Cares,* both illustrated by Rigby Graham, and Arthur Rimbaud's *Les Corbeaux,* illustrated with woodcuts by Robert Blythe. Rigby Graham has recorded that of the copies of Clare's *Lines* 'scores were offered for next to nothing in pubs, used as beer mats, and eventually hurled at sneering crowds in the market place on Saturday morning.' The Tichbourne pamphlet contained the haunting poem which he wrote the night before he was executed for his part in the Babington Plot.

During this time Douglas Martin had moved to Munich with the intention of restarting the press in a much more lavish way. He never achieved this ambition but returned from Germany and once more the press resumed activity. On his return he managed through the patronage of a wealthy businessman to buy machinery and trays of type. He moved into his first proper premises at 140 London Road, Leicester, and within six months had printed *Kirby Hall* and published John Best's *Poems and drawings in mud time.*

Kirby Hall: an uninformative guide, 1960, was printed one afternoon and taken from the press to be circulated in the local pub – The Hind. Ten copies of this booklet were printed on special paper made for it at the Holt Mill, Aylestone. Forty copies were printed on white mould-made paper, although of these many were faulty and so the actual number of completed copies was nearer thirty. The book was actually published from the Island of Elba. The press had by this time run out of money and the last publication, *Poems and drawings in mud time,* was printed at the Curwen Press with John Best paying for its publication.

The imprint was revised once by Rigby Graham, Patricia Green and Toni Savage with Thomas Churchyard's *Lovesong to an Inconstant Lady,* 1961. They subsequently adopted the name Pandora for future publications.

The Orpheus Press had a short but eventful life. The press had moved about Leicester and travelled abroad to publish books in Bavaria, Switzerland and Paris. Many ambitious projects were started, including Herrick's *Hesperides,* Shakespeare's *Sir Thomas More,* and a *Life* of the composer, Bellini amongst others.

So chaotic an undertaking as Orpheus was unlikely to last, and according to Douglas Martin was not intended to. He finds it of interest that Orpheus has apparently served as a stimulus to so much excellent private press printing in Leicestershire. Since Orpheus he has been art editor for Hodder & Stoughton's children's books, design consultant to a number of other publishers, and (administratively) involved with the Biennale of Illustrations Bratislava and the International Book Art Exhibition Leipzig.

les corbeaux

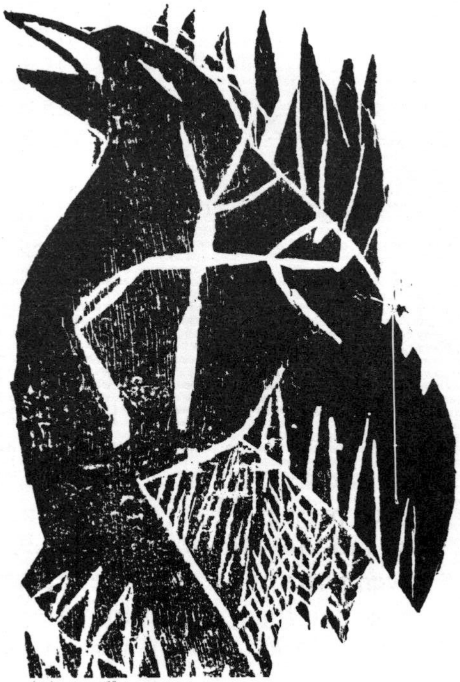

Seigneur, quand froide est la prairie,
Quand dans les hameaux abattus,
Les longs angelus se sont tus . . .
Sur la nature defleurie
Faites s'abattre des grands cieux
Les chers corbeaux delicieux.

Armee etrange aux cris severes,
Les vents froids attaquent vos nids!
Vous, le long des fleuves jaunis,
Sur les routes aux vieux calvaires,
Sur les fosses et sur les trous
Dispersez-vous, ralliez-vous!

Par milliers, sur les champs de France,
Ou dorment les morts d'avant-hier,
Tournoyez, n'est-ce pas, l'hiver,
Pour que chaque passant repense!
Sois donc le crieur du devoir,
O notre funebre oiseau noir!

Mais, saints du ciel, en haut du chene,
Mat perdu dans le soir charme,
Laissez les fauvettes de mai
Pour ceux qu'au fond du bois enchaine,
Dans l'herbe d'ou l'on ne peut fuir,
La defaite sans avenir.

Orpheus Press. *Les Corbeaux.* Illustration by Robert Blythe

Saint Anthony Press

The last press to be established in the 1950s was the Saint Anthony Press. The press was run by C. P. Healey, who brought to private printing a very different view of life because, not only had he spent several years in National Service, but he had also been a tramp! He became a student at the Leicester College in his late twenties and he became interested in printing and bought an Albion press, which was formerly owned by Dr Outram of the Rosemary Press. He paid £3.10s for it and transported it via a wheelbarrow to his home, and the Saint Anthony Press began operating. In 1959 he produced a poem by Oliver Bayldon called *Morning.* In 1959 he also produced Ray Seaton's *Russia* with illustrations by Rigby Graham. The other productions are *Sol-Mari,* 1960, written by C. P. Healey himself, *My Island Star,* 1960, by H. I. Bergstrom, and *Llanto por Ignacio Sánchez Mejías* by Federico García Lorca, 1962. This last production was rather different from the previous productions which had been quite small. It was the largest book from the press and was translated into English by A. L. Lloyd. It included fine woodcuts by Rigby Graham.

Pandora Press

The 1960s could be called the decade of the private press in Leicestershire because most of the presses which are in existence today were established during this time. The first press to be formed in 1960 was the Pandora Press.

Rigby Graham, Patricia Green and Toni Savage were the people responsible for the establishment of the Pandora Press. Before establishing Pandora the three partners had produced *Lovesong to an Inconstant Lady,* 1961, under the Orpheus imprint, but subsequently adopted the name Pandora. After the Orpheus Press had ended they thought that, as in Pandora's box, all that was left was hope. Therefore Hope became their press mark, a strange winged animal with a snout – a cross between a flying pig and a printer's devil.

The equipment they used was limited and consisted of an Adana, a nipping press, two wooden presses and a small amount of type. The partners still managed to produce a considerable number of books rapidly – in the first seven years of its existence it produced over twenty titles. The press had no fixed abode, for a time it operated in an attic of an old rectory at Aylestone, then in a loft in the village of Ashby Magna near Rugby. Yet another time it appeared from Walton Street, Leicester.

Patricia Green, Rigby Graham's wife, set almost all the type, Toni Savage did the printing with Rigby Graham illustrating the books and pamphlets. They decided to print only what they wanted to print and although they did not aim at a profit, costs could not be forgotten because at that time none of the partners was exactly affluent. They wanted the illustrations to be many and varied and the books and pamphlets were to be the largest they could manage – the press took a maximum of eight by ten inches. Their aim was to 'lose money slowly' but they rarely managed to get enough money back to cover blocks and materials. The choice of material was by personal preference governed by the amount of type available.

The first Pandora publication was in 1961 by Edward de Vere, the Earl of Oxford, called *The Sheepheard's Commendation of his Nymph.* This was followed rapidly by an essay of John Clare's, *The Natural World,* 1961, Algernon Charles Swinburne's *The Garden of Proserpine,* 1961, and Shelley's *Lines written among the Euganean Hills,* 1961, – their most ambitious early publication. All are set in Baskerville type and most effectively bound in elephant hide. The book by Shelley was bound in Continental style with panelled boards and contained an introductory note by Thea Scott. All four books were illustrated with line drawings by Rigby Graham and printed in a variety of colours.

In 1962 they produced seven publications. Most of these were small pamphlets such as Algernon Charles Swinburne's *A Match,* Andrew Marvell's *Thoughts in a Garden,* and Byron's *When We Two Parted.* All are very well designed and printed. Most of the Pandora productions have illustrations in colour calling for two or three impressions.

One Pandora publication, *The Living Theatre,* 1962, was particularly interesting. This was an account of a struggling professional theatre company in Leicester. The tiny theatre was faced with increasing financial difficulties which finally overcame them. Its struggle, collapse and subsequent publicity were instrumental in the eventual foundation of the Phoenix Theatre.

Two more ambitious and on the whole very successful books were Thea Scott's essay on the topography and wild life of Staffa, *Fingal's Cave,* 1962, and Penelope Holt's essay, *A Sicilian Memory,* 1963.

On the most substantial book of the press, *Fingal's Cave,* the 52 pages were so large, 10 × 8½ inches, that they had to be printed one at a time and side-stitched into the blue cloth boards. Disaster struck the project after the seventy-sixth printing: at the time Pandora was operating from Rigby Graham's home, an old rectory at Aylestone, and the sheets had been spread out on the floor to dry. A sudden spell of sunshine hatched out woodworm in the floor and overnight more than one hundred of the sheets were peppered with holes. The edition was

THE MAN OF LIFE UPRIGHT
THOMAS CAMPION
illustrated by Rigby Graham

Pandora Press : 1962

Original printed in black and dark blue

therefore reduced to 150 copies. Both *Fingal's Cave* and *A Sicilian Memory* were bound by Trevor Hickman.

They later produced *Rigby Graham: an introduction to his work as an artist* which was published by the Gadsby Gallery when it held an exhibition of his work. *Nine Gnats,* 1964, by John Best and Rigby Graham, was a small booklet containing correspondence, poems and three drawings by Rigby Graham to fit in with the poems. Previously, John Best and Rigby Graham had collaborated on *Poems and drawings in mud time,* 1960, published by the Orpheus Press.

In 1965 the Pandora Press produced *Stille Nacht, Heilige Nacht,* a note on the carol by Patricia Green. The next year they printed something else for Christmas, a letter from *Johann Wolfgang von Goethe.*

The last production was *Poems and Translations,* 1967, by Count Potocki of Montalk.

In addition to these books, the Pandora Press also printed *Fishpaste.* It was thought to be the smallest little magazine in Europe as it was printed on a postcard. Each issue was to include a poem or some critical prose cut down to the minimum and illustrated with linocuts, drawings, or collages. *Fishpaste* developed into a lighthearted project, although it had started quite seriously. Peter Hoy, then a lecturer in French at Leicester University, had a collection of French little magazines. They were published by a variety of people and covered a multitude of subjects in various languages. Pandora Press decided such a magazine could be printed on anything – even a postcard. It was called *Fishpaste,* according to Rigby Graham, because nobody knew what would be in it! The series was intended to run to 21 numbers and sell at one guinea for the complete set or a shilling each. The actual numbers issued were *Fishpaste* 1–22, plus title page, *Fishpaste* 4½, 6a, Umpteen, the special announcement and hors-serie 1, 1 and 3, and a colophon. These were printed at Pandora between January 1967 and August 1968.

In addition, some forgeries were openly known to exist. Roderick Cave of the Library School, Loughborough, for instance, issued *Fishpaste* 34. Originally the intention was to print an edition of one hundred but it was soon evident this was too small and so they doubled the edition – they never ended up with so many as there was considerable wastage. *Fishpaste* had a change of address mid-way through its production because Peter Hoy moved to Oxford. The cards were still printed in Leicester but at first inadvertently printed the address as Holywell Road and not Street which was Peter Hoy's new home. Gradually the number of *Fishpaste* increased from the original 21 but the practical participation of Pandora ended with *Fishpaste* 22. Peter Hoy and Rigby Graham later produced one further edition of *Fishpaste* in November 1968 in honour of Guillaume Apollinaire and an unnumbered extra edition was produced. Peter Hoy did start a second series of

Fishpaste and eight issues were printed at the Daedalus Press in Stoke Ferry. *Fishpaste* had been the biggest and most adventurous job Pandora had undertaken.

The Pandora Press ended about 1968. It had an eventful life and had produced very lively and interesting work. Like Orpheus it experienced financial difficulties. For the participants it was only a temporary setback and Rigby Graham and Toni Savage were later to establish their own presses.

Percival and Graham

When Rigby Graham was still a partner in the Pandora Press he became involved in the production of two beautiful books which were printed by the monks of Mount Saint Bernard Abbey. These were produced with George Percival, the binder of many of the Stanbrook Abbey publications. The first of these productions was *Vale,* 1961. It was printed as a farewell to John Mason on his retirement from the College of Art. The book gave a humorous account of John Mason's career and was printed deliberately haphazardly with phrases upside down, lines of capitals and other such tricks. It was printed on Twelve by Eight paper purloined (they claimed) from John Mason. It was bound in Chinese style with a dazzling puce front cover board. The book was presented to John Mason at a commemorative dinner held in June 1961. The other book was equally interesting, this was Oscar Wilde's *The Nightingale and the Rose,* 1961. It was bound with aquamarine green metallic boards and pink metallic satin doublures blocked in gold. The subject of the book was the rather sad tale of the Nightingale who died in his attempt to provide a red rose for an intending young lover. It was a beautiful example of book production.

It is worthwhile to note here the work of the Cistercian monks of Mount Saint Bernard Abbey. The Mount Saint Bernard Press is at Whitwick, in Charnwood Forest, and was established in 1954. They are basically jobbing printers who print business cards, letterheads, leaflets and commercial work of all kinds. Besides printing the two books mentioned earlier, they have also worked with the Brewhouse Press of Trevor Hickman printing the text and illustrations for *An Autumn Anthology,* 1965.

Brewhouse Press

Of the Leicestershire private presses, the Brewhouse Press has produced the most substantial books. The Brewhouse Press is at Wymondham, a

司祭叙品記念

一九五九・三・十四
於ローマ
パウロ 高橋重幸

John Mason disguised as the eunuch TSai Lun, 106 AD

He started life as a compositor but very quickly realised the limitations of this following and swiftly

Percival and Graham. *Vale John Mason.* Drawing by Rigby Graham, in black and dark green on blue paper little village close to the border with the former county of Rutland. The owner of the press, Trevor Hickman, is a lecturer in bookbinding and one of the few people in the private press scene who has actually trained and worked in the commercial book trade. He was trained as a bookbinder and was encouraged to become a part-time teacher. In 1955 he again came into contact with Rigby Graham, whom he had last met as a student in 1949. Through the influence of Graham, Trevor Hickman developed a keen interest in private publishing. One of the results of their friendship was that Trevor Hickman bound for the Pandora Press *Fingal's Cave* and Penelope Holt's *A Sicilian Memory.* Much of the work issued by the Pandora Press was in pamphlet form which did not provide a great enough challenge for a bookbinder of his calibre; therefore, encouraged by Graham, the Brewhouse Press was started in 1962.

The press takes its name from its home, an ancient brewhouse attached to Trevor Hickman's house. Hops were grown in the area, and still continue to grow in the hedgerows, so a clump of hops became the Brewhouse pressmark. As the equipment was limited, it was decided that Brewhouse would undertake only the work best done by itself. The processes carried out more easily by friends and associates of Hickman were to be done elsewhere. As a result, much of the printing of texts is done at Melton Mowbray, Coalville and by the Daedalus Press at Stoke Ferry. The illustrations are also printed in these places and at Walton-under-Edge. Brewhouse is responsible for the editing, typography, binding, gold-stamping and the majority of the experimental prints, as well as printing shorter pieces of text.

Trevor Hickman enjoys binding his books in unusual materials. He finds that by mastering new problems and experimenting with new materials he is able to pass on this knowledge to his students. His aim is that each publication will be as different as possible and bear no resemblance to the previous ones. The text is of great importance – it must be original and not a rehash of others' work. Editions have been lavish and limited; one or two of the publications have run to an edition of 400 copies but Hickman believes that his maximum will be 450, as beyond that he thinks the press will be similar to a commercial undertaking.

In 1964, the first production issued was *Chidioch Tichbourne,* a twelve page booklet, containing a letter on his life, his elegy, and a poem written by Tichbourne to his wife the night before he was executed for his complicity in the Babington conspiracy. The booklet reflected something of the style of production of the Pandora press and for Trevor Hickman this was a disaster – he wanted books and not pamphlets.

The next title was *An Autumn Anthology,* 1966, which was a collection of poetry and prose, printed by the monks of Mount

This book is a collection of prints which resulted from sketches and ideas inspired by Pickworth – its ruins and its associations. It is not complete nor can it ever be. Pickworth is but a fragment and this little book is a series of fragments.

Brewhouse Press. *The Pickworth Fragment.* Drawing by Rigby Graham

St Bernard's Abbey. Two editions were issued, an ordinary and a special; a practice which Brewhouse adopted for many of its productions.

In 1966 Brewhouse produced an unusual book, Rigby Graham's *The Pickworth Fragment.* Pickworth was a village in Rutland, destroyed in the War of the Roses and later ravaged by plague. John Clare, the poet, worked there as a lime-burner, and his poem on Pickworth is included in the book. The book is unusual because the text and brightly coloured lino-cuts were printed on one side only, joined and folded to make a concertina nearly 14 ft long; a successful variation on a Japanese form of binding. The whole is contained in a quarter-binding of brown suede with gold-tooled and coloured linocuts on the brown paper boards. The book was selected for the National Book League's exhibition of British books in 1966, and subsequently exhibited at Frankfurt in 1967.

The next production in 1967 was *Typographia Naturalis* by Roderick Cave and Geoffrey Wakeman, which gave a detailed account of the history of nature printing. It included the original description by Leonardo da Vinci, together with an account of the work of Henry Bradbury in the nineteenth century, and some interesting examples of contemporary work.

In 1968 came *A Musical Breakfast* by Philip Ward, Brewhouse's first step into drama. A science fiction play in two acts was illustrated with lithographs by Rigby Graham; one lithograph combined lithography and gold-stamping, a technique which is rarely used as a method of book illustration. The same year saw the production of *The Oakham Canal* by David Tew. This is a history of the canal from its inception to its closure and beyond. Rigby Graham and Trevor Hickman walked the canal's route from Melton Mowbray to Oakham in order to make the account as accurate as possible. David Tew, a solicitor and friend, was responsible for finding the facts about the canal. Today, it is regarded as the authoritative work on this canal and it is hoped that a second edition will be produced.

In September 1968, Brewhouse produced Peter Hoy's *Camus in English* which was an annotated bibliography of Albert Camus's contributions to periodicals and newspapers. The list included extracts from Camus's fiction, essays, and letters as well as the more significant authenticated interviews.

Gold and Books by Penelope Holt and Edmond Thorpe was produced in 1969. This was a history of the ways in which gold has been employed in book production from the earliest times. Appropriately the book was bound in gold skivertex with full gilt edges and a gold-blocked title page spread. The cover includes an exceptional example of gold stamping using an interlocking design in two colour gold. A very scholarly book was produced in 1970; this was Geoffrey Wakeman's *Aspects of Victorian lithography, anastatic printing and photozincography.*

Typographia
Naturalis

by
RODERICK CAVE
and
GEOFFREY WAKEMAN

BREWHOUSE PRESS
WYMONDHAM
1967

Original in black and gold

It was issued in an edition limited to 250 copies and bound in quarter black morocco with marbled paper covered boards. *An Alderney Afternoon* which followed it was a very different book. It was an account of the Brewhouse jaunt to the Island of Alderney.

Rigby Graham and Trevor Hickman originally went to Guernsey to help organise the Book Fair and decided to visit Alderney for just half a day. They borrowed some bicycles and cycled around the island. The island with its remains of German fortifications made a deep impression on both of them and when they returned to Wymondham it took only fourteen days to produce the book.

Prizefighting at Wymondham in 1811 between Tom Cribb and Tom Molyneux was the subject of a book produced in 1971 entitled *Tom Cribb at Thistleton Gap.* It was written by Dennis Prestidge and illustrated by Rigby Graham. The fight took place in the field between Thistleton and Wymondham on 28 September 1811. It was the first defence for the heavyweight championship of the world. The site was chosen because it is in a unique situation where three counties meet and very near to the Great North Road. Another Brewhouse jaunt was to the Lake District one Easter in order to illustrate *Mountains,* a poem on the area by Mervyn James. This proved to be a very successful production.

The most ambitious book they attempted in 1972 was *The Casquets* written by Rigby Graham. This is a beautifully printed and bound book limited to thirty copies. It contains a wealth of Rigby Graham's drawings of the Casquets, the very treacherous rocks near Guernsey. *The Casquets* is a perfectionist exercise in printing on high quality paper. It has marbled Ortac endpapers especially produced for this edition by Douglas Cockerell. Each copy bears a metal block of the Casquets against rich red black on its blue-grey cloth covered front board.

The Brewhouse Broadsheets began in 1966 and were published during the succeeding years until the last one appeared in 1974. Each of the thirteen broadsheets consisted of a french folder contained within a wrapper, on varying subjects.

The first was Trevor Hickman's *The execution of a bookbinder.* The others have included *Paul Gauguin, woodcutter and private printer,* and *Monotypes* by Rigby Graham. *Monotypes* is one of the series of Brewhouse publications on the history and practice of printing and print-makers. It includes a large monotype of Kildownet Castle.

Silence at Midnight and *Silence à Minuit* contained in the same folder constituted Broadsheets 3 and 4. This was an account of clandestine printing in France during the Nazi occupation. The Broadsheets have now been discontinued. Brewhouse Broadsheet number 13 is the last and contains a potted history of the press. Trevor Hickman has always been more interested in books than pamphlets and feels that the broadsheets have reached their natural conclusion.

Besides producing its own books and broadsheets, Brewhouse has bound books for others and has produced special bindings of its own publications. It has bound, for example, the Orpheus Press production of *Die Sonnette an Orpheus.* Another aspect of Brewhouse has been the organisation and staging of exhibitions both in Leicestershire and abroad.

Two catalogues which Brewhouse produced for exhibitions are *The Private Press Today* by Juliet Standing produced for the King's Lynn Festival of 1967, and *A Bibliography of Private Publishing* by G. Stephens-Cox, 1970. This was published from Guernsey and was both a catalogue to the Guernsey Book Fair and a bibliography of the Toucan Press of Guernsey and the Brewhouse Press – the title was misleading in this respect. The Brewhouse has also been involved with exhibitions of private press work at Watford, the Chameleon exhibition at King Street, Leicester, Guernsey, Glasgow and in America with an exhibition of private press work at Clarksville, Tennessee.

In 1974, Brewhouse published a book on Peddars Way in Norfolk, the route Rigby Graham, David Rogers and Trevor Hickman walked one summer. Like *The Pickworth Fragment* this book represented Great Britain at the Frankfurt Book Fair.

Another production which is planned is a translation of Charles Baudelaire which he is doing in conjunction with the monks of Mount St Bernard Abbey. It is to be printed on hand-made paper – his first book on such paper – and will be bound in some unusual material. The Brewhouse press illustrates how a private press can experiment with new materials and processes. The books often involve a considerable amount of hand work – something which no commercial press would consider economically practical.

Cog Press

Rigby Graham dominates the private press scene in Leicestershire because he has been involved in so many presses. These include Pandora, Percival and Graham, Threoteotha, and more recently Wind Tunnel Irregulars and the Group One Press. He has also contributed through his artistic work to a large number of publications of private presses and little magazines elsewhere in Britain and abroad. Although trained as a mural painter, he has achieved considerable success as a painter, illustrator, graphic designer, writer and bookbinder. His work in the field of graphics is becoming particularly well known and it is difficult to realise how he manages to find the time for the work he contributes to the private press field.

Cog Press. *Cogs in Transition* by Rigby Graham. Original in black, ochre and red

The Cog Press operates from Rigby Graham's home and was founded in 1963 as a small experimental press. Rigby Graham intended Cog to be a lighthearted project and it has never possessed any definite objectives. The Cog Press began because of Rigby Graham's involvement with the Private Libraries Association's Society of Private Printers. This organization consisted of a number of people offering to print and willing to exchange their publications with others. Rigby Graham printed his first Cog book for this society, called *Cogs in Transition.* The press takes its name from the cog wheel which he took from an alarm clock. The equipment used is an Adana HQ hand press and a nipping press. He also obtained an ancient 'Favourite' gas-operated gold stamping press, previously used in the shoe industry, now converted to natural gas.

The press produces books and pamphlets and illustrates them with prints from natural and man-made objects. In 1965 it produced *El Icaro* by Salvador Jacinto Polo de Medina, which was translated by Patricia Green. A year later came a catalogue with notes and a preface by Peter Hoy entitled *An exhibition of graphic art from the French little magazines.* This exhibition was held in Leicester from November 1966 until January 1967. It was the first of several catalogues of exhibitions. The others were: *John Minton: a commemorative exhibition,* 1967; *Catalogue to an exhibition of woodcuts,* 1969; and *Little magazines: from the collections of Rigby Graham and Alec Morris.* This last book was a catalogue to an exhibition of little magazines held in Leicester in 1970. All the catalogues were sumptuously bound in limited editions.

The press also produces *Enigma.* This is devoted to literature and graphics, and includes essays, poems and prints by Rigby Graham and others. So far *Enigma 4* has been reached and 5 and 6 are planned for the future.

In 1968 he produced *Five Duets* at the request of the author, Hugh Collinson. Subsequently one or two were quarter bound in continental style in turquoise blue oasis morocco. One notable book produced by Cog in 1968 is *Slieve Bingian: a cycle of prints and drawings.* It was written and illustrated by Rigby Graham and is one of the books he greatly enjoyed doing. Although he enjoys producing the books at the time, very few Cog or any other books have produced any lasting pleasure for him. Typically, he regards none as his best work — presumably that is to be achieved in the future.

Two books produced recently are *Cobwebs from California* by Leon Spiro and *St Julian* by Melinda Camber. *Cobwebs from California* is illustrated by Rigby Graham and printed on mould-made paper. It is the first book of Leon Spiro's to be published. He lives in California and has been writing since 1963 and publishing other poets internationally.

The Cog Press has produced about twenty broadsides which have been given away to friends. It has also printed one or two of the Phoenix

Enigma 3

edited by Rigby Graham

COG PRESS
1970

Broadsheets for Toni Savage, and half a dozen Barwell Broadsides distributed free at the Ampersand Folk Club at Barwell near Hinckley. These, like the Pandora issues of *Fishpaste* were not listed in *Private Press Books.*

Rigby Graham has also illustrated and printed material for other publications, private presses and little magazines. Cog produced many illustrations for the poetry magazine *Priapus,* which was run by John Cotton. When *Priapus* ended in 1972, Cog produced the cover of a small edition of *Priapus* especially for the Poetry Society's farewell to the magazine. *Cog* also produced John Cotton's volume, *Ampurias* (Priapus Press, vol 1). Rigby Graham in addition has contributed illustrations for the cover and title pages of *In Particular 1,* and for *The Private Printer and the Private Press.*

Black Knight Press

The Black Knight Press was established in 1964 by Duine Campbell. At first the press operated from 6 Lansdown Place West, Bath, but then Duine Campbell moved to Leicester. In 1967 the press was established at 6 Connaught Street, its present address.

Duine Campbell is an art teacher in Wigston, who actually took a degree in English at Newcastle. The first press he ever owned was the Grand Piano (one linocut under a grand piano) and his second press had an equally idiotic name, the Trouser Press.

The name of the Black Knight press is taken from the symbolic order of knights, the Black Knights being the lowest in the order of chivalry. As a beginner in the printing world, with very little knowledge, it seemed an appropriate title. Duine Campbell sees the objective of the press as bringing people together – books as the means of introducing an illustrator to a writer through a printer. Together they can explore the possibilities of their collaboration in book form. This kind of partnership is not always possible in the commercial world.

The equipment of the press is an Adana HQ and since 1971 he has used occasionally a Thomson Number Two Gem treadle platen press.

Black Knight productions have generally been small books in limited editions, printed on English or Japanese hand-made papers. Recently, larger and more lavish books have appeared. The press is also responsible for many ephemera, such as tickets, invitations to exhibitions, Christmas cards and so on, all of which display the great care and consideration given to the presswork.

The second Black Knight production from Bath was *Till the Leaves Come* by Vic Atkins. This was followed by *A story,* 1966, a gruesome legend by the same author. Penelope Campbell, Duine Campbell's sister, illustrated *Two Poems,* 1968, by Francis Ponge, which were printed on Japanese Hoso-shi paper.

Duine Campbell finds that Blake has a perpetual appeal for him and has frequently used Blake's texts as an inspiration for his own illustrations. The books produced include *A Divine Image,* 1968, *Infant Sorrow,* 1970, and *Proverbs of Hell,* 1972. *Infant Sorrow* included three poems by Blake illustrated with four linocuts and printed in three colours. *Proverbs of Hell* was printed on Japanese hand-made papers and contained an extract from Blake's *The Marriage of Heaven and Hell* with an introduction by Jack Schofield, the owner of the Offcuff Press, British Columbia.

In 1969 *An Old Drama: three encounters with Jack the Ripper* was produced; a haunting tale illustrated with line drawings by Deirdre Maton. *After the game,* written by Pat Smith, which was published in 1971, is the reminiscences of the dying victim of an atomic attack during the Third World War. The illustrations are by Lynn Taylor.

A more ambitious project was *The Rubaiyat of Omar Khayyam,* printed in two editions. The special edition, priced at £18, was bound by Alan Constance of Bristol, in full blue oasis morocco. Printing was completed in 1970, but binding delayed publication until 1971. Each page was printed separately and over 25,000 impressions were required for the two editions.

In 1973, Black Knight Press produced a similarly lavish production of Idris Parry's *Stream and Rock.* This was a revised and extended version of a BBC Radio 3 talk given by Idris Parry, Professor of German Literature at Manchester University. Twenty copies were bound by Sangorski and Sutcliffe, in full grey oasis morocco, with a slip-case. The other 80 copies were quarter bound in grey morocco with Ingres paper-covered boards.

In preparation is a portfolio using the text of *The Marriage of Heaven and Hell* by Blake. This is a joint effort with members of staff at Bushloe High School and it will be issued free in an edition of approximately 75 copies. Another work in progress is *The Butterflies are here;* a short story in verse by Professor William Bradley of Pennsylvania State University. The line block illustrations are by Maria Winkler Green. An edition of 100 copies is envisaged, to be specially bound by Robert Green of London. The equipment of 39 Connaught Street is also used by the Offcut Press and by the New Broom Press.

I feared the fury of my wind
Would blight all blossoms fair & true;
And my sun it shined & shined
And my wind it never blew.

But a blossom fair or true
Was not found on any tree;
For all blossoms grew & grew
Fruitless, false, tho' fair to see.

Black Knight Press. *Infant Sorrow.* Drawing by Duine Campbell

Threoteotha Press

The Threoteotha Press has also been in existence since 1965. It has produced a considerable number of fairly substantial books since its establishment. Unlike other presses, it is not the work of a fixed group of people but the people involved varies with each edition. The guiding light behind its work is Rigby Graham. Threoteotha is the Anglo-Saxon word meaning thirteen, and although the number involved has not always been thirteen, the name seems appropriate to a small group of people, some of whom have been students of Rigby Graham. The equipment they used was generally an Adana flatbed handpress and binder's nipping presses, but in 1968 they started using an Albion press for some books. The editions are usually limited to from twelve to fifteen copies; one is given to each person who has contributed to the book. The books have never been published or offered for sale. Trevor Hickman has been responsible for the binding of some of the editions.

In 1965 Threoteotha produced *Havelock Street,* which contained twelve prints from lino and other surfaces by thirteen artists. The subject of the book was a street in Leicester which was about to be demolished and they recorded in the prints impressions of this dying street. In 1965 they also produced *A Print Book of Arches and Bridges in and around Leicester.* This consisted of 37 lino cuts by twelve artists and a lino block reproduction of a pen drawing.

Pattern of Brickwork followed in 1965 with Rigby Graham and twelve other artists contributing to its production. In this book patterns made by brick were taken as the theme and developed in a variety of ways.

Two years later they produced *People: a book of prints,* with Rigby Graham and seventeen other artists taking part in it. Each book had a collection of contributions either written or drawn and this idea was continued in the books which followed such as *Leicester,* 1968, and *Imbroglio,* 1968.

Some Notes on Woodcuts by Rigby Graham was also produced in 1968. It had photo-litho and line block reproductions of seven woodcuts and wood engravings by Dürer and others, and included fifteen woodcuts by Rigby Graham and ten other artists. The same year also saw the production of *Machines: a book of prints.*

Polymorph '69 followed the next year, and in 1970 they produced *To have and to hold,* a book where the words from the marriage ceremony were used to entitle and unite a varied collection of prints and drawings. *Polyphonic Vision,* 1971, *English woodcuts,* 1972, and *Minority luggage,* 1973, were the titles of later productions from the press which follow the same pattern as previous books.

The very latest production is called *Integer,* 1974, which was

Threoteotha Press. *Pattern of Brickwork.* Illustration by Thomas Frame produced during the energy crisis in January, and February 1974 by a small group of technical teachers from colleges in Leicestershire and Northamptonshire.

Offcut Press

In 1967 The Offcut Press was established as a result of collaboration between Patricia Harris, Sheila Cafferky, Duine Campbell and Toni Savage. Their aim was to produce books printed on offcuts, as well as to have fun, and drink coffee. The press is run from the same address as the Black Knight Press, using the same equipment, and in fact wore out the Adana but purchased a new one for the use of both presses. As most of the publications are printed on offcuts, the size is small. The first production was called *Plus or minus half an inch,* 1967, 'wrung from' Sheila Cafferky. This little book contained illustrations by Duine Campbell and a few lines of verse.

Frontispiece drawn by Duine Campbell. Original in black and red on toned paper

The same year came the production of *My Life Halted* by William Moore Powell and *A curious collection of remedies: taken from a Book of Notable Things,* by Lupton. Subsequent productions include *Pastorale,* 1968, by Count Potocki of Montalk, George Peele's *Twice so fair,* 1970, and *La Barcillunisia,* a traditional Sicilian poem translated by Toni Savage.

In 1969 they produced a little book, *Excerpts from Offcut,* which was presented to guests at the private view of the exhibition of private press books mounted by Duine Campbell and Toni Savage at The Chameleon, King Street, Leicester on the 2 February 1969. Afterwards the pressmen still able to walk went to 'The Flying Horse' (where James Cook the bookbinder met John Paas in 1832 before murdering him). The book contained extracts from Offcut's past publications and a few original items. As the Offcut Press continued, it became the responsibility of just Toni Savage and Duine Campbell.

One of the interesting booklets from Offcut was *Values* by Spike Milligan, 1969, which understandably went out of print quickly. One of the least successful booklets was *Floppy the Jujiburra*, 1971, a children's story by John Pearse. It was for sale at £6 but has never been in great demand. The same year they produced *Miscellany for Daisy,* an interesting little booklet (containing poetry and prose by William Shakespeare, W. B. Yeats and others) for their most faithful customer, Daisy Lane, on her retirement to a round-the-world cruise.

Two imprints useful to note here in connection with the Offcut are the *Offcuff Press* and the *New Vacuum Cleaner Press.* The Offcuff originally operated from Yorkshire but then Jack Schofield moved to Vancouver. It seems that Jack Schofield visited Connaught Street and produced and printed *Sweet Basil,* 1970, and *Happy Families,* 1970. Duine Campbell and Toni Savage were also responsible for the New Vacuum Cleaner Press which produced one item, *Liniment: a lament for Lim Meng Lin.*

When Douglas Martin, formerly owner of the Orpheus Press, visited Connaught Street, he printed *The Revenge of Orpheus* which contained a little verse:

'Offcut, Pandora, St Anthony & the Twelve by 8
figs as measure and lack of time dictate –
follow the pub-crawl which Orpheus erst did make;
a printer's thirst with cut-price sheets to slake.'

Plough Press

The Plough Press was founded in Loughborough in 1967 by Geoffrey Wakeman who had learned to print from Philip Gaskell at the College Press in Glasgow. The first book printed by the press was *Share of ploughs,* 1968, which consisted of a series of coloured blocks of ploughs with accompanying verses about ploughs and ploughing. The owner's apparent obsession with ploughs was simply due to a desire to utilize some old electros depicting them which had been brought to Loughborough from Trinidad by Roderick Cave. The name of the press was derived from the same collection.

In 1967 a small book, *Anaglyptography,* had been printed for the press by Tom Rae, the first work to appear under his new imprint of Grian-aig. This contained an inserted specimen of anaglyptography and was an early example of this press's subsequent devotion to the production of leaf books. The first important example of this genre was *XIX century illustration* written by the printer and consisting of about 23 specimens of different illustration methods with accompanying letterpress, the whole contents arranged in 15 sewn fascicules. The text was handset in 18 pt Bembo with the assistance of Rod Cave and John Walton, the former also making the boxes in which everything was contained. It was printed on a late nineteenth century Golding Pearl treadle press which had been bought for the purpose.

In addition to large works like *XIX century illustration,* the press has produced some jeux d'esprit: in 1971 *Loughborough marble* appeared, purporting to give examples of a local paper marbling industry. In 1973, *Impressions of binder's tools,* demonstrated the press finding how difficult it is to get a succession of good impressions from a collection of binding tools which had recently been acquired.

By and large, it is one of the objectives of the press to produce books which could only be made by hand, but occasionally this is lost sight of. A straightforward reprint of an account of papermaking from an 1835 *Book of trades* turned out to involve a lot of tedious typesetting without any particular success in the finished product. A more significant book was *English hand made papers suitable for bookwork,* which was optimistically dated 1972 on the title page but suffered delays in binding which prevented its actual appearance before 1973. This consisted of all the different varieties of hand-made paper up to 150 gsm, which could be readily bought in 1972. The samples were printed with appropriate texts or illustrations of the various processes of making paper by hand.

In 1974 and 1975 most of the efforts of the press were being put into a successor to *XIX century illustration,* dealing with colour printing in British book illustration. This dealt with the most important methods of printing in colour and the printers who used them. It had 17 pages of text, each facing an original colour print. As many of the prints in any one part of the book had to have different captions, depending on where they came from, and all had to be inserted by hand, this book was well within the objectives of the press.

ENGLISH HAND MADE
PAPERS

SUITABLE FOR BOOKWORK

Assembled by Geoffrey Wakeman

THE PLOUGH PRESS
1972

New Broom Press

Toni Savage started the New Broom Press in 1968. Duine Campbell went away on holiday and Toni Savage took advantage of this unique opportunity while keeping an eye on the house to print two books and one pamphlet – all under the imprint New Broom. Toni Savage has led a varied life; he was once a lyric tenor at the Teatro la Fenice in Venice and has been involved in various Leicester presses. He helped for a time at the Orpheus Press and later ran the Pandora Press with Patricia Green and Rigby Graham. The New Broom Press operates from Toni Savage's home in Cambridge Street, Leicester. He is a partner in the Offcut Press and sometimes helps with the presswork for Black Knight. Since 1968 the New Broom Press has produced a large number of booklets of merit, which are generally quite small with paper covers.

The two books which Toni Savage produced in ten days while Duine Campbell was on holiday were the following: *Spring,* poems by Peter Blythe with hand-coloured illustrations by Zeta Alessandra (Toni Savage's daughter); followed rapidly by *Canzone,* an anthology of poems chosen by Toni Savage. He also produced a pamphlet by Chris Bartram. After that, New Broom continued producing at a fairly rapid pace. In 1968 he also produced *Cicadas* by Count Potocki of Montalk and for Christmas he printed *The time draws near the birth of Christ* by Tennyson. In 1970 he produced *Lammas Day* by Count Potocki of Montalk decorated with hand coloured lino cuts by Rob Armstrong.

One of the best of the earlier booklets was *Lapsus Calami* by Boyd K. Lichfield, with lively drawings by Franco Colavecchia. In the same year, 1969, he produced a tiny book containing a short poem on the disaster in the Welsh village of Aberfan. It was entitled *Children of Aberfan* and was written by Spike Milligan. Under the offcut imprint Toni Savage also printed with Duine Campbell's wood engravings Leslie Shepard's *Seascape,* which the author described as 'a pattern of sounds for reading aloud.'

Poetry and verse have predominated in the work of New Broom. It has printed poems by Shelley, Keats, Byron and more contemporary verse unpublished elsewhere by such people as Eric Swift, Jake Thackray, William Powell, Colin Scot, Peter Blythe and Don Partridge. In 1973 New Broom published *Manifesto* by Brian Allen, *The Still,* by Don Partridge, famous for his one-man-band act, and *Bitter Taste,* poems by Colin Scot, the contemporary folk and blues singer.

The most substantial volume to date has been Rex Brisland's *Canticle Collection,* 1971, printed on turquoise green paper with purple endpapers and a blue binding by Terry Sharrott. It contained a mixture of poems and song lyrics. This is not only the work of Brisland's that the New Broom produced. The year before it had published his *Triptych* and the year after came his *Fairy Story. Canticle Collection* and *Fairy Story* contained coloured illustrations by Marlene Staniforth. A later (and in Toni Savage's opinion his best) work is *The Lost Shepherd,* 1974, by Frank Lissauer.

The content of the New Broom books has generally been obtained from Toni Savage's friends. The standard of illustration has tended to be mixed, but that of printing has always been high. The culmination of Toni Savage's theatrical and folk song interests has been the production of a series of Broadsheets. Since 1972 New Broom has become increasingly well known for these Broadsheets many of which are distributed free of charge to the audience of the Phoenix Theatre, Leicester. Toni Savage has been interested in this type of literature for a long time and has made a collection of street literature, theatre bills and various broadsheets.

Whilst he was helping with the Orpheus Press, he persuaded Douglas Martin to produce Broadsheets although at that time only one was ever issued – *Vaga luna che margenti* – a little Italian song by Vincenzo Bellini. He was also closely involved with *Fishpaste* and the broadsheets reflect the influence of these former years. The Phoenix Broadsheets have proved to be very successful and are eagerly sought after. Each broadsheet contains a poem, translation or short piece of text and an illustration. Some of the writers of the Broadsheets are friends of Toni Savage, texts have included fragments or short works by Thomas Carew, William Barnes, Boyd Lichfield, Spike Milligan, Don Partridge, Jack Woolgar, and many others. They have been illustrated by a wide variety of artists including Rigby Graham, Marlene Staniforth and Brian Allen.

It is fitting that Toni Savage with his theatrical background should become involved with the professional theatre again with the broadsides. Their actual beginning was in 1972 at the Poetry and Folk Week held at the Phoenix Theatre, Leicester. Some of the sheets have been produced for special occasions such as the premiere of J. B. Priestley's 'Open House', and for a special celebration of the magazine *Priapus* held at the Poetry Centre, Earls Court. The sheets have been produced on a variety of papers varying from commercial printers' offcuts to hand-made. Toni Savage has also produced a few broadsheets for other theatres in Leicester, the Haymarket and the Little Theatre. It is noteworthy that despite the eagerness with which the Broadsheets are sought, Toni Savage still produces them free of charge.

A booklet, *Muinestruck* by Angela Thigmold, was printed by Toni Savage and Patricia Harris. It contained puns on the name 'Duine' and was illustrated by wood engravings 'stolen' from Duine Campbell.

In 1969 Zeta Alessandra and Gillian Greaves, both then thirteen

years old, printed two of their poems in a booklet entitled *One and one*. It was produced under the imprint of the Steven Press – derived from the name of Zeta Alessandra's cat. They used the equipment at 39 Connaught Street, printing the booklet themselves with minimal help from Toni Savage. Other productions were *Desiderata* and a booklet of poems about Toni Savage and his drunken pals.
Zeta Alessandra also helped with the hand colouring of many of the Phoenix Broadsheets.

Phœnix Broadsheet No. 22

LOOK, LOVE

Now that the days of rest are gone
(a thunderbolt brushing the farmer's cheeks)
and the cries of the children squeezed
between the fireworks of our feast
are reaching us,
look, love......
look......
look......
The horizon is misty,
No ship of hope can sail into our harbours,
The wind's rebelling
and the foggy sea's around
like heavy breathing pounding
in a chilly old man's breast.
look, love......
look......
look......
But let's forget all and become sterile
of these filthy thoughts:
Your body's mine under these sheets,
Mine is yours......
Let's bear a little poet......
Despite the weariness tomorrow.

Charles Flores

Toni Savage of Leicester. December 1972

Original in grey on lemon paper

Gazebo Press

This was established in 1965 by Michael Felmingham, a librarian. When he was living in Birmingham Michael Felmingham was involved with a fellow librarian, Geoffrey Beard, and Arthur Knowles in the so-called *Post Boy Press.* Productions from that press are said to be rare and according to some non-existent. Michael Felmingham moved to Leicester to become a librarian at the College of Art. He seems to have been a person with a wide knowledge and interests. He knew a considerable amount about type which is not altogether surprising in view of his career. However, what is unusual is that he obtained a teaching post in the printing school. The Gazebo Press was operated from Huncote and in 1965 he produced the first book, using a small Adana press. This was *Alphabet* with illustrations provided by Michael Felmingham himself. Thirty copies were printed and four copies were bound by Trevor Hickman. In 1966 he produced his second and last work from Leicestershire. This was Loys Le Roy's *The Smoke and Sweat of Oyle,* a contemporary account of Renaissance printing. After that, he moved away from Leicestershire and he now lives at Leamington Spa in Warwickshire.

Medbourne Press

This was the second press to emerge in 1965. It operated from Hinckley Road, Leicester, and was run by John Lythgoe. Only one booklet was produced from that address, Charles Pickering's *Three Poems.* He used an Adana hand press and the occasion for the production was to celebrate the author's seventy-fifth birthday. The booklet was not for sale but 100 copies were printed, 50 for the printer and 50 for the author. John Lythgoe later moved to Romanby Northallerton in Yorkshire.

Silk Purse Press

This is a little-known press which operated from Melton Mowbray. The press made its appearance when L. F. Taylor, the owner, brought it to the Private Press Exhibition, held at King Street, Leicester. The printing press caused a considerable amount of comment because the bed of the press was made of glass. In 1968, *A Useful Titbit* was produced, and also a Christmas card with a long series of lino cuts for the twelve days of Christmas.

Wind Tunnel Irregulars

This is the name of another little press which operates from Leicester, largely inspired by Rigby Graham. The name of the press derives from the long windy passage that leads from Pocklington's Walk to Newarke Street in Leicester. In 1969, a group of print-makers produced *Newarke Street* – a poem by Chris Bartram, illustrated with twenty-two experimental prints from lino and other surfaces by Rigby Graham and other artists. The Adana press belonging to the Cog Press was used, together with various nipping presses. Only fifteen copies were printed, none of them for sale.

Transican Press

Boyd Lichfield ran the diminuitive Transican Press from a thatched cottage at Newtown Linford, on the edge of Charnwood Forest. Lichfield taught himself the rudiments of printing when he set up a press in an old house in Kimberley Road, Leicester. He is a poet, who started the press in order to print his own work and that of his friends. His poetry has been broadcast by the BBC and one of his poems appeared in the *Phoenix* Broadsheet No. 1. One book printed entirely on a variety of Japanese hand-made papers was *Walking Out,* a leather-bound volume of the early poems of Brian Patten. This was followed by *The Dissolving Cuckoo,* and in 1970 by *Old Man in a Garret* – poems exploring 'an octogenarian's life whilst his faculties are still sensitive to the world about him'. Twenty-three copies were printed, but the Gartree Press in Leicester has reprinted it in a larger edition.

Group One Press

Another little known press is the Group One Press. Only one book has been published, *Creare,* 1972, by Michael Smith, who was a drummer in a jazz group in Leicester. The book contained poems, articles, a variety of prints, lino cuts, transfer prints and a 'pop-up'. Fifteen copies were lithographed, typeset and printed by Patricia Green and John Tyson.

POLLUTION

Smoke and fumes fill the air
Dirt and grime gets in our hair
Children playing in the street
Paper flying round their feet.

The river full of tin and can
All this mess was caused by man
The river once was clean and bright
Now its just a muddy sight.

If only man would stop and think
And take a bit more care
This world of ours would be so clean
And full of good fresh air.

Elaine Smith

Crypt Slip One 1973

Crypt Press Broadside

Crypt Press

Another press operating in the early seventies was The Crypt Press of Terry St Clair. He produced *You didn't need me* with drawings by Brian Allen, in a limited edition of one hundred copies. *Crypt Slip One* is a broadsheet, similar to the *Phoenix Broadsheets,* produced in 1973, consisting of a poem, *Pollution* by Elaine Smith.

Apple Barrel Press

Gordon Atkins, senior design lecturer at Leicester School of Printing, established the Apple Barrel Press towards the end of 1972, having at one time been involved with Triad Publications (with John Mason).

The Merry Tales of the Wise Men of Gotham was produced in 1973. Gordon Atkins has taken the old tales by Andrew Boorde, and illustrated them. The book is set in Baskerville type and illustrated with lithographs, silk-screen and lino-prints. It includes a reproduction of a 17th century woodcut, a Bewick engraving and a double spread of the original book in the Bodleian Library, Oxford. *The classification of printing types* appeared in 1975, written, designed and produced by Gordon Atkins. It presents an account of the history of type classification from the first attempt of Theodore Low de Vinne to the present. It is unusual for a private press book in being photo-set and printed by offset lithography.

Ashby Lane Press

The Ashby Lane Press was started by Bryan Foster going to evening classes at Leicester in 1968 to learn typesetting. His first use of the imprint was on a programme for GUS Footwear Band in June 1970. He started printing on an old office letter copying press, then graduated to a Jardine Treadle Platen which was in its turn replaced by a 15 × 11 Victoria Platen. He then bought from an old Coalville printer a Royal (26 × 20) Britannia, serial number 205. It is undated but thought to be from the eighteen thirties. He printed mainly posters and other ephemera until 1974 when *Six Poems* by Sue Barraclough appeared. This was followed in 1975 by Baron Munchausen's *The Woollen Horse,* lavishly illustrated with woodcuts by James Deacon, and signed by the author. This is Bitteswell Chapbook Number 1.

Ashby Lane Press. *The Woollen Horse.* Woodcut by James Deacon

The Private Presses and their publications

APPLE BARREL PRESS
Gordon Atkins, 20 Woodfield Road
Oadby, Leicestershire

Gordon Atkins
The Merry Tales of the Wise Men of Gotham
1973, 25 pp, 100 copies

Gordon Atkins
The Classification of Printing Types
1975, 64 pp, 100 copies

ASHBY LANE PRESS
Bryan Foster, Bitteswell
Near Lutterworth, Leicestershire

Sue Barraclough
Six Poems
1974, 14 pp, 120 copies

Baron Munchausen
The Woollen Horse
Illustrated by James Deacon
1975, 12 pp

BLACK KNIGHT PRESS
Duine Campbell, 39 Connaught Street
Leicester

Jim Roebuck
The Poppy's Song
Illustrated by Deirdre Carr
Bath, 1965, 7 pp, 90 copies

Vic Atkins
Till the Leaves Come
Illustrated by Jean Cecil
Bath, 1966, 18 leaves, 70 copies

Vic Atkins
A Story
Illustrated by Pete Harper
1966, 10 pp, 100 copies

Pat & Ros Smith
Next Monday Happening
Illustrated and printed by the authors
1967, 2 pp, 55 copies

Francis Ponge
Two Prose Poems
Translated by Peter Hoy
Illustrated by Penelope Campbell
1968, 18 pp, 120 copies

William Blake
A Divine Image
Illustrated by Duine Campbell
1968, 13 pp, 100 copies

Duine Campbell (ed)
Dream Interpretations: a selection from the book of notable things
1969, 16 pp, 100 copies

Michael Gardner
An Old Drama: Three Encounters with Jack the Ripper
Illustrated by Deirdre Maton
1969, 27 pp, 60 copies

William Blake
Infant Sorrow
Illustrated by Duine Campbell
1970, 9 pp, 100 copies

Edward Fitzgerald (trans)
The Rubaiyat of Omar Khayyam
Illustrated by Steven Morris
1971, 67 pp, 250 copies

J. Wesley
Receipts from Primitive Physic
Linocuts by Andy Hughes
1971, 20 copies

Duine Campbell
Catalogue of Books Printed at the Black Knight Private Press
1971, 8 pp, 140 copies

Hilaire Belloc
The Microbe
Illustrated by Lim Meng Hooi
1971, 35 copies

Pat Smith
After the Game
Illustrated by Lynn Taylor
1971, 19 pp, 100 copies

Leonardo da Vinci
Precepts of the Painter
Illustrated by Philip Fongod
1972, 25 copies

William Blake
Proverbs of Hell
Illustrated by Duine Campbell
1972, 28 pp, 100 copies

Duine Campbell
Black Knight Private Press – Catalogue of Books in Print
1972, 8 pp, 100 copies

Priory Road, 1972–3
A portfolio of printed things (in a sense)
A cooperative effort by members of the staff and students of the School of Art Education, Birmingham
1973

Idris Parry
Stream and Rock
Illustrated by Islwyn Watkins
1973, 29 pp, 100 copies

Duine Campbell
Books in Print at the Black Knight Press
1973, 10 pp, about 150 copies

BROADSHEETS:

1
William Blake
Till a System was Formed
1972, 35 copies

2
William Blake
London
1973, 35 copies

3
Art and the Military Mind
Wood engraving by Duine Campbell
1973, 35 copies

BREWHOUSE PRESS
Trevor Hickman, The Orchard
Wymondham, Leics

Chidiock Tichbourne
Chidiock Tichbourne: his last letter to Agnes his wife, his elegy together with an answer to Mr Tichbourne and a biographical note
Illustrated by Rigby Graham
1964, 12 pp, 250 copies

Pat Graham (ed)
An Autumn Anthology
Illustrated by Rigby Graham
1965, 22 pp, 140 copies

Rigby Graham
A note on the book illustrations of Paul Nash
With illustrations by Paul Nash
1965, 31 pp, 155 copies

Rigby Graham
The Pickworth Fragment
Illustrated by Rigby Graham
1966, 32 pp, 75 copies

Gadsby Gallery Exhibition
Paintings by Rigby Graham: Sculpture by K. Ford
1966, 8 pp

John Best and Rigby Graham
Poems and Drawings in Mud Time
Illustrated by Rigby Graham
1960 *Orpheus Press* (qv)
Brewhouse took four copies, added three extra illustrations and bound them in full leather.

Roderick Cave and Geoffrey Wakeman
Typographia Naturalis: a history of nature printing
1967, viii, 39 pp, 300 copies

Juliet Standing
The Private Press Today
Two editions: 24 pp and 32 pp
1967, 375 copies. Produced for the Kings Lynn Festival, 1967.

Philip Ward
A Musical Breakfast: a one act play in two scenes
Illustrated by Rigby Graham
1968, 20 pp, 325 copies

David Tew
The Oakham Canal
Illustrated by Rigby Graham
1968, 119 pp, 450 copies

Peter Hoy
Camus in English
Illustrated by Rigby Graham
1968, 31 pp, 250 copies

Hugh Collinson and Rigby Graham
Five Duets
The ordinary edition of 71 copies was produced at the Cog Press. The special edition was a joint Cog Brewhouse production.
1968, 18 pp, 4 copies

Warren Taylor
Rigby Graham – Paintings
Catalogue to an exhibition of paintings at the Gadsby Gallery, Leicester
1969, 12 pp, 100 copies

Penelope Holt and Edmond Thorpe
Gold and Books
Illustrated by Rigby Graham
1969, 33 pp, 250 copies

August Stramm
August Stramm: twenty-two poems
Illustrated by Rigby Graham
1969, 23 pp, 200 copies

Patrick Bridgwater
Rigby Graham
Catalogue to an exhibition of paintings held at the Griffon Garnett Galleries, Shrewsbury
1969, 9 pp, 100 copies

Geoffrey Wakeman
Aspects of Victorian lithography, anastatic printing and photozincography
1970, 64 pp, 250 copies

Rigby Graham
An Alderney Afternoon
Illustrated by Rigby Graham
1970, 52 pp, 30 copies

Stephen Stevens-Cox
A Bibliography of Private Publishing
1970, 40 pp + poster folded and tipped in, 200 copies

Dennis Prestidge
Tom Cribb at Thistleton Gap
Illustrated by Rigby Graham
1971, 68 pp, 450 copies

Mervyn James
Mountains
Illustrated by Rigby Graham
1972, 46 pp, 226 copies

Rigby Graham
The Casquets: the most dangerous Channel Islands
1973, 91 pp, 30 copies

David Rogers
The Peddars Way
Illustrated by Rigby Graham and Trevor Hickman
1974, 98 pp, 100 copies

Rigby Graham
Edmund Spenser's Kilcolman
1975, 24 pp, 80 copies

Thomas Hood
Faithless Nelly Gray
Illustrated by Rigby Graham
1975, 12 pp, 80 copies

David Tew
Sewstern Lane
Calligraphy by Maureen Hallahan
Illustrated by Jill York
1975, 36 pp, 100 copies

BREWHOUSE BROADSHEETS:

These consisted of a single french-folded sheet giving 8 pages size $10\frac{1}{2} \times 7\frac{5}{8}$ inches when folded.

1
Trevor Hickman
The execution of a bookbinder, James Cook
Illustrated by Rigby Graham
1966, 75 copies

2
Roderick Cave
The Dance of Death
Illustrated by Trevor Hickman
1967, 150 copies

3 & 4
Peter Hoy
Silence at Midnight and Silence à Minuit
Illustrated by Rigby Graham
1967, 150 copies

5
Rigby Graham and Trevor Hickman
One hour less, a bookbinder dies from gaol fever
1968, 200 copies

6
Penelope Holt
Roger Payne, bookbinder
Illustrated by Rigby Graham
with reproductions of Payne's work
1969, 200 copies

7
Rigby Graham
Paul Gaugin: woodcutter and private printer
Illustrated with reproductions of Gaugin's work
1969, 150 copies

8
Count Potocki of Montalk
Myself as a Printer
Illustrated by Rigby Graham
1970, 240 copies

9
Rigby Graham
Monotypes
Illustrated by Marilyn Andreeti and Michael Ayrton
1970, 100 copies

10
John Lord
William Blake: private printer
1972, 300 copies

11
Alan Tucker
Reading Games
Nine plates
1973, 300 copies

12
Gregory Stevens-Cox
The earliest books printed in Guernsey
1973, 300 copies

13
Patricia Graham
The Brewhouse at Wymondham
Illustrated by Rigby Graham
1974, 90 copies

COG PRESS
Rigby Graham, 396 Aylestone Road
Leicester

Rigby Graham
Cogs in Transition: selections from a mechanical sketchbook
Illustrated by Rigby Graham
1963, 16 pp, about 60 copies

Jacinto Polo de Medina Salvador
El Icaro: The Icarus
Illustrated by Rigby Graham
1965, 7 pp, 40 copies

Peter Hoy
An exhibition of graphic art from the French little magazines
1966, 51 leaves, 20 copies

Rigby Graham
John Minton: a commemorative exhibition
1967, 110 pp, 40 copies

Hugh Collinson
Five duets
Illustrated by Rigby Graham
1968, 15 pp, 75 copies

Rigby Graham
Slieve Bingian: a cycle of prints and drawings
Illustrated by Rigby Graham
1968, 56 pp, 48 copies

Rigby Graham
Catalogue to an exhibition of woodcuts
1969, 45 pp, 20 copies

Rigby Graham and John Cotton and others
Enigma 1
1969, 18 leaves, 60 copies

Rigby Graham (ed)
Enigma 2
1970, 32 leaves, 50 copies

Little Magazines: from the collections of Rigby Graham and Alec Morris
Catalogue to an exhibition held at Leicester
1970, 56 leaves, 20 copies

Rigby Graham (ed)
Enigma 3
1970, 50 leaves, 50 copies

Rigby Graham (ed)
Enigma 4
1971, 64 leaves, 50 copies

Charles Flores
Correspondence
Illustrated by Rigby Graham
1973, 10 pp, 150 copies

Charles Flores
Memories
2-colour linocut
1973, 20 copies. A Cog Press Broadsheet

Leon Spiro
Cobwebs from California
Illustrated by Rigby Graham
1973, 20 pp, 40 copies

Melinda Camber
St Julian
Illustrated by Rigby Graham
1973, 6 pp, 35 copies

COG PRESS BROADSIDES

1973 issues

1
Leon Spiro
Flight Time
Drawings by Rigby Graham

2
Charles Flores
Memories
Linocut by Rigby Graham

1974 issues

3
George Crabbe at Stathern
Illustration by Rigby Graham

4
Leon Spiro
St Mary's Nether Broughton
Illustrated by Rigby Graham

5
Grainnes Castle – Achill Island to County Mayo
Illustrated by Rigby Graham

1975 issues

Pro Bono Publico
Cog broadsheet produced for the Come All Ye Exhibition at Glasgow

7
George Borrow
A Prayer and a Letter
Illustrated by Rigby Graham
Cog sheet produced for the Print Your Own Thing lecture at Glasgow School of Art

8
Humbert Wolfe
Bethlem
Produced for the Print Your Own Thing Lecture at Glasgow School of Art

9
Rigby Graham
Kildownet Castle
Illustrated by Rigby Graham

10
Spailpín a Rúin
Plastic engraving by Rigby Graham

11
A Rough Night
Illustration by W. H. Bartlett

12
David G. Rogers
Samuel Johnson at Market Bosworth
Illustration from Nichols *History of Leicestershire*

13
The Remains of Edmund Spenser's Kilcolman Castle near Donevaile
Illustration by Rigby Graham

14
Sceilig Michael
Illustration by Rigby Graham

15, 16 & 17
An exhibition of street literature and private press broadsheets
Illustrated by Rigby Graham

18
The Cog Press
with a photograph by Tony Martinez. A french fold broadsheet printed for the Society of Private Printers

19
Cadeby Light Railway

PHOENIX BROADSHEETS:

20
43 *Irish Proverbs*
Illustrated by Rigby Graham 1975

21
46 *More Irish Proverbs*
Illustrated by Rigby Graham 1975

BARWELL BROADSIDES:

22
1 James Clarence Mangan
I have pleasure in nothing
Drawing by Rigby Graham

2 *Three Epitaphs*
Drawing by Rigby Graham

3 Emily Bronte
Fall Leaves
Drawing by Rigby Graham

4 *The wind blows high*
Drawing by Rigby Graham

5 David G. Rogers
Samuel Johnson and the Battle of Bosworth with 2 illustrations from John Nichols *History of Leicestershire*

6 Henry King
Of Human Life
Drawing by Rigby Graham

23
J. M. Synge
Prelude
Drawing by Rigby Graham

24
Both Eddie and Finbar Furey, Jennifer Gentle with a drawing by Rigby Graham (there were two issues with different drawings)

CRYPT PRESS
Terry St. Clair
11 Sunnydale Road, Hinckley
Leicestershire

T. St. Clair
You didn't read me
Illustrated by Brian Allen
1972, 14 pp, 100 copies

GARSWOOD PRESS

A. E. Christopherson
The Wisdom of Andrew Boorde
Illustrated and designed by author
1936, 63 pp, 550 copies

Falconer Scott
Crowns May Perish
Illustrated by 'S. H.' 2 pp
Setting designed by T. S. Haile and the programme by Stanley Hickson

Aria Da Capo
A play in one act
Cover designed by Stanley Hickson
1936, 2 pp
The play was produced by A. E. Christopherson

GAZEBO PRESS
Formerly 30 St James Close
Huncote, Leicestershire

Michael Felmingham
Alphabet
Illustrated by Michael Felmingham
1965, 31 pp, 30 copies

Loys Le Roy
The smoke and sweat of oyle
Illustrated by Michael Felmingham
1966, 12 pp, 60 copies

GROUP ONE PRESS

Michael Smith (ed)
Create
Various illustrators
1972, 86 pp

MEDBOURNE PRESS
Formerly 432 Hinckley Road
Leicester

Charles Pickering
Three Poems
1965, 12 pp, 100 copies

NEW BROOM PRESS
Toni Savage, 78 Cambridge Street
Leicester

Peter Blythe
Spring
1968, 10 pp, 50 copies

Chris Bartram
Poem
Linocut by Chris Bartram. 1968

Toni Savage (ed)
Canzone
1968, 14 pp, 100 copies

Count Potocki of Montalk
Cicadas
With linocuts by Bill Powell
1968, 14 pp, 80 copies

Alfred Lord Tennyson
The time draws near the birth of Christ
Illustrated by Duine Campbell
1968, 10 pp, 50 copies

Count Potocki of Montalk
Lammas Day
1969, 26 pp, 100 copies

Geoff Smedley
Geoff's Medley
Illustrated by Rob Armstrong
1969, 14 pp, 70 copies

Boyd K. Lichfield
Lapsus Calami
Illustrated by Franco Colavecchia
1969, 19 pp, 100 copies

Spike Milligan
Children of Aberfan
1969, 6 pp, 30 copies

Percy Bysshe Shelley
Chorus to Hellas
Illustrated by Rod Felton
1969, 10 pp, 40 copies

Eric Swift
Nine Poems
Illustrated by Marlene Staniforth
1970, 18 pp, 110 copies

Stan Kelly
Four Blue Gargles
Illustrated by Marlene Staniforth
1970, 11 pp, 50 copies

John Lyle
Cupid and Campaspe
Illustrated by Marlene Staniforth
1970, 7 pp, 40 copies

Toni Savage
An exhibition of street literature
From the collection of John Foreman,
Eric Swift and Toni Savage
1970, 48 pp

Andrew Marvell
The Fair Singer
Illustrated by Amy Dibl
1970, 6 pp, 70 copies

Jack Harris
The Battle of Leon: an unofficial souvenir
Illustrated by Zeta Alessandra
1970, 8 pp, 250 copies

Rex Brisland
Triptych
Illustrated by Zeta Alessandra
1970, 9 pp, 120 copies

Robert Burns
Winter
Illustrated by Marlene Staniforth
1970, 8 pp, 50 copies

Colin Scot
April in July
Illustrated by Rod Felton
1971, 18 pp, 130 copies

Colin Scot
Three Flowers
Illustrated by Marlene Staniforth
1971, 10 pp, 100 copies

Colin Scot
Idle Scratches
Illustrated by Brian Allen
1971, 11 pp, 140 copies

Peter Blythe
The Light
Illustrated by Brian Allen
1971, 8 pp, 80 copies

Rex Brisland
Canticle Collection
Illustrated by Marlene Staniforth
1971, 46 pp, 130 copies

Boyd K. Lichfield
The Phoenix
Illustrated by Brian Allen
1971, 6 pp, 100 copies

Frank Lissauer
Elementary lament
Illustrated by Brian Allen
1972, 40 copies

Finbar Furey
Until We Meet
Illustrated by Marlene Staniforth
1972, 50 copies

Rex Brisland
Fairy Story
1972, 11 pp, 100 copies

Brian Allen
Manifesto
Screen printed illustrations by Brian Allen
1973, 24 pp, 100 copies

Don Partridge
The Still
Illustrated by Marlene Staniforth
1973, 8 pp, 130 copies

Colin Scot
Bitter Taste
Illustrated by Chris Bartram
1973, 12 pp, 120 copies

Jack Woolgar
Nostalgia
Illustrated by Marlene Staniforth
1973, 60 copies

Frank Lissauer
The Lost Shepherd
Illustrated by Rigby Graham
1974, 125 copies

Charles Flores
A Voice from Kalkara
Illustrated by Rigby Graham
1974, 103 copies

Ralph Waldo Emerson
The Snowstorm
Drawings by Rigby Graham
1974, 8 pp

Jack Woolgar
Triolets
Illustrated by Kathie Layfield
1975, 20 pp, 125 copies

John Foreman
Blackboard of Misfortune
Illustrated by Kathie Layfield
1975, 24 pp, 125 copies

Jack Woolgar
Christmas
Illustrated by old woodengravings of sailing ships. 1975

PHOENIX BROADSHEETS

1971

1
Boyd K. Lichfield
For Brian Patten
Illustrated by Duine Campbell

2
R. Armstrong
Acquiescent
Illustrated by E. Whymper

3
Brian Patten
The Projectionist's Nightmare
Illustrated by Brian Allen

4
Rex Brisland
Shadow
Illustrated by Marlene Staniforth

1972

5
William M. Powell
We Go Down

6
Colin Scot
Summer Incident
Illustrated by Marlene Staniforth

7
Steven Swindon
Rain
Illustrated by Marlene Staniforth

8
Thomas Frazer
Walking Home
Illustrated by Chris Bartram

9
William Barnes
A Winter Night
Illustrated by Duine Campbell

10
Don Partridge
Reflection upon Reflection
Illustrated by Toni Savage

11
Thomas Frazer
Iron Giant
Illustrated by Rigby Graham

12
John Betmead
Love Song
Illustrated by Marlene Staniforth

13
The Words of Finn
Illustrated by Toni Savage

14
Don Partridge
Public Toilet
Illustrated by Pete Bogge

15
Thomas Carew
Good Councel to a Young Maid
Illustrated by Marlene Staniforth

16
Padraig Shanahan
Dun Aengus-Irishmore
Illustrated by Rigby Graham

17
Geoffrey Shovelton
West Meadow Park
Illustrated by Tony O'Dwyer

18
Colin Scot
The Final Act
Illustrated by Brian Allen

19
George Tebbutt
Boswell Tries to Leave Corsica
Illustrated by Rigby Graham

20
William Smyth
No More, My Mary, I Sigh for Splendour
Illustrated by Rigby Graham

21
A Farewell to Priapus
Illustrated by Rigby Graham

22
Charles Flores
Look Love
Illustrated by Franco Colavecchia

1973

23
Eddie Furey
Dawning of the Day
Illustrated by Marlene Staniforth

24
Christine Coburn
Humanity
Illustrated by Brian Allen

25
Colin Scot
The Broken Dish

26
Jack Woolgar
Rondeau
Illustrated by Duine Campbell

27
William Shakespeare
O Mistress Mine
Illustrated by Marlene Staniforth

28
John G. Pearson
What's the Use

29
Boyd K. Lichfield
Mermaid, Mouse and Map
Illustrated by Chris Bartram

30
Jake Thackray
The Suckling
Illustrated by Rob Armstrong

31
David G. Rogers
The Recruiting Officer
Illustrated by Rigby Graham

32
Petruchio's speech from 'The Taming of the Shrew'
Illustrated by Brian Allen

1974

33
Geoffrey Shovelton
Meadfoot Nocturne for Solo Tenor

34
R. E. Turner
She
Illustrated by Rigby Graham

34 [bis]
Henry James
Ruins. English Hours
1905. Illustrated by Rigby Graham

35
Rex Brisland
Circle of Life
Illustrated by Duine Campbell

36
Jack Woolgar
The Theatre
Illustrated by Rigby Graham

37
Hans Erni Exhibition
Illustrated by Hans Erni

37 [bis]
Frank Lissauer
Song: in memorie of Hylton Small of Leycestre

1975

38
John Foreman
Love
Illustrated by Rigby Graham

39
Bill Gadsby
A. H. Findley
Illustrated by Marlene Staniforth

40
Finbar Furey
A star winks
Illustrated by Rita

41
Erik Ilott
Two views of life at sea
Illustrated with an old wood engraving

42
Sion Powell
Cockrow
Illustrated by Duine Campbell

43
Irish Proverbs
Illustrated by Rigby Graham

44
Jack Woolgar
Seasons
Illustrated by Marlene Staniforth

45
Count Potocki of Montalk
White Furtive Forms
Illustrated by Duine Campbell

46
More Irish Proverbs
Illustrated by Rigby Graham

47
The Dance
Illustrated by Marlene Staniforth

48
Music Hall
Illustrated by Rita

49
Barry Skinner
Tones of Life
Illustrated by Brian Allen

50
Padraig Shanahan
Janus
Illustrated by Rigby Graham

51
Victor Fenech
A Smile
Illustrated by Thomas Bewick

52
Jack Woolgar
Triolet – Nightmare
Illustrated by Duine Campbell

53
Dominic Timms
Solitude
Illustrated by Rod Felton

53 [bis]
Jeremy Sandars
Silences
Illustrated by Chris Bartram

54
Barry Skinner
Age
Illustrated by Duine Campbell

55
Jack Woolgar
Hedge-Pig
Illustrated from an original block by Thomas Bewick

56
Ted Furey
The Angelus
Illustrated by Tony O'Dwyer

57
Cliff Aungier
Lost
Illustrated by Rod Felton

58
Jenni Trevitt
I'll be your Friend
Illustrated with an old wood cut

59
William Hazlitt
Johnny Boxer
Illustrated by Rigby Graham

60
Alex Campbell
Life
Illustrated by Rod Felton

61
Spike Milligan
Rose Petals Taut
Illustrated by Rigby Graham

62
Frank Lissauer
Her Seasons
Illustrations by Barbara Jones cut by P. P. Piech

OFFCUT PRESS
Toni Savage and Duine Campbell
39 Connaught Street, Leicester

An Epitaph
1967, 1 pp
A keepsake

Revenge of Orpheus – a poem
1967, 2 pp, 20 copies

Sheila Caffeky
Plus or Minus Half an Inch
Illustrated by Duine Campbell
1967, 16 pp, 60 copies

William Moore Powell
My Legs Halted
Illustrated by Tony O'Dwyer
1967, 15 pp, 100 copies

A Collection of Remedies: taken from a book of notable things
Illustrated by Duine Campbell
1967, 14 pp, 100 copies

Toni Savage
Butterfly Antics
Illustrations by Duine Campbell 1967

Patrick Smith
Ice Queen
Illustrated by Duine Campbell
1967, approx 50 copies

La Barcillunisia
Translated by Toni Savage
Illustrated by Duine Campbell
1968, 8 pp, Approx 100 copies

Count Potocki of Montalk
Pastorale
Illustrated by Tony O'Dwyer
1968, 14 pp, 120 copies

Percy Bysshe Shelley
Italian Idylls
Illustrated by Duine Campbell
1968, 180 pp, 20 copies

Rita Forman
Hilaritas
Illustrated by Rita Forman
1968, 18 pp, 100 copies

Bill Powell
Song of the Armed Angel
Illustrated by Bill Powell
1968, 2 pp, 50 copies

Sir Philip Sidney
With How Sad Steps
1968, 2 pp, 100 copies

Excerpts from Offcut
1969, 15 pp, 75 copies

Spike Milligan
Values
Illustrated by Rigby Graham
1969, 23 pp, 150 copies

Rex Brisland, Jack Harris, Mark Newman
The Couriers
1969, 6 pp

Abigail Fawcitt
Pottets
Illustrated by Rita Foreman
1969, 15 pp, 80 copies

Mike Jacobs
Uncollected Poems
Illustrated by Rob Armstrong
1970, 18 pp, 150 copies

George Peele
Twice so Fair
Illustrated by Paul Peter Piech
1970, 15 pp, 200 copies

Vera Johnson
Flamingo
1970, 20 copies

Shirley Toulson
All Right Auden, I Know You're There
Illustrated by 'A. J. E.' and 'D. C.'
1970, 7 pp, 75 signed copies

John Pearse
Floppy the Jujiburra
Illustrated by Marlene Staniforth
1971, 19 pp, 170 copies

Jack Schofield
The Midas Touch
Illustrated by Duine Campbell
1971, 15 pp, 100 copies

The Song of Solomon
Illustrated by Duine Campbell
1971, 2 pp, 15 copies

William Shakespeare and others
Miscellany for Daisy
Illustrated by Duine Campbell
1971, 14 pp, 80 copies

Leslie Shepard
Seascape
Illustrated by Duine Campbell
1971, 5 pp, 130 copies

Jean Roberts
Henrik Ibsen's 'Peer Gynt'
1972, 10 pp, 75 copies

ORPHEUS PRESS
Douglas Martin, 69 Stoughton Drive
Oadby, Leicester

John Clare
Lines written in Northampton County Asylum
Illustrated by Rigby Graham
1959, 8 pp, 70 copies
Second edition of 100 copies with added bibliographical note

Rainer Maria Rilke
Die Sonette an Orpheus
Illustrated by Rigby Graham
1959, 66 pp, 25 copies

Arthur Rimbaud
Les Corbeaux
Illustrated by Robert Blythe
1959, 8 pp, 50 copies

Chidioch Tichbourne
My Prime of Youth is but a Frost of Cares
1959, 8 pp, 145 copies

Douglas Martin
Kirby Hall
Illustrated by Rigby Graham
1960, 12 pp, about 50 copies

John Best
Poems and Drawings in Mud Time
Illustrated by Rigby Graham
1960, 48 pp, 1000 copies, only 500 of which were bound

Thomas Churchyard
Lovesong to an Inconstant Lady
Illustrated by Rigby Graham
1961, 8 pp, 220 copies

PANDORA PRESS
Formerly c/o Toni Savage
48 Walton Street, Leicester

Edward De Vere, Earl of Oxford
The Sheepheard's Commendation of his Nymph
Illustrated by Rigby Graham
1961, 13 pp, 140 copies

John Clare
The Natural World
Illustrated by Rigby Graham
1961, 8 pp, 40 copies

Algernon Charles Swinburne
The Garden of Proserpine
Illustrated by Rigby Graham
1961, 16 pp, 125 copies

Percy Bysshe Shelley
Lines written among the Euganean Hills
Illustrated by Rigby Graham
1961, 48 pp, 60 copies

Algernon Charles Swinburne
A Match
Illustrations by Rigby Graham
1962, 12 pp, 150 copies

George Gordon, Lord Byron
When We Two Parted
Illustrated by Rigby Graham
1962, 12 pp, 100 copies

The Living Theatre
Illustrated by Rigby Graham
1962, 16 pp, 250 copies. Sold for and by the 'Living Theatre'

Thomas Campion
The Man of Life Upright
Illustrated by Rigby Graham
1962, 12 pp, 130 copies

Andrew Marvell
Thoughts in a Garden
Illustrated by Rigby Graham
1962, 12 pp, 120 copies

Oscar Wilde
Serenade
Illustrated by Rigby Graham
1962, 12 pp, 100 copies

Thea Scott
Fingal's Cave
Illustrated by Rigby Graham
1962 (but dated 1961), 51 pp, 150 copies

A. W.
Where His Lady Keeps His Hart
Illustrated by Rigby Graham
1963, 12 pp, 70 copies

Penelope Holt
A Sicilian Memory
Illustrated by Rigby Graham
1963, 33 pp, 70 copies

Toni Savage
Rigby Graham
With a mural painting note by Hugh Collinson
1963, 25 pp, 100 copies

John Best
Nine Gnats
Illustrated by Rigby Graham
1964, 15 pp, 50 copies

Epitaph for Jones: an inscription from a headstone in a churchyard in the county of Norfolk
Illustrated by Rigby Graham
1965, 11 pp, 50 copies

Patricia Green
Stille Nacht, Heilige Nacht
Illustrated by Rigby Graham
1965, 7 pp, 50 copies

A letter from Johan Wolfgang von Goethe, 25th December 1787
Illustrated by Rigby Graham
1966, 8 pp, 50 copies

Count Potocki of Montalk
Poems and Translations
Illustrated by Rigby Graham
1967 (dated 1966), 24 pp, 50 copies

PERCIVAL AND GRAHAM

Vale
Illustrated by Rigby Graham
1961, 8 pp, 20 copies

Oscar Wilde
The Nightingale and the Rose
Illustrated by Rigby Graham
1961, 44 pp
Both printed by the monks of Mount Saint Bernard Abbey, and bound by George Percival and Rigby Graham.

PLOUGH PRESS
Geoffrey Wakeman
28 Pantain Road, Loughborough
Leicestershire

William John Stannard
Anaglyptography
1967, ii, 11 pp + plate, 200 copies

Share of Ploughs
1968, 22 pp, 35 copies

*Law and ordure **1969***
1969, 12 pp, 40 copies

Geoffrey Wakeman
XIX Century illustration, some methods used in English books
1970, 7 pp, prelims + 15 sewn fascicules, 75 copies

Geoffrey Wakeman
Loughborough marble
1971, 9 pp, 18 copies

The Papermaker, reprinted from the Book of Trades (1835)
1971, 15 pp + tipped in frontispiece, 60 copies

English hand made papers suitable for bookwork
1972, 39 leaves, 75 copies

Impressions from binders tools
1973, 12 leaves, 12 copies

Geoffrey Wakeman
Plates to accompany 'Victorian Book Illustration'.
1974, iii, 20 plates + facing captions, 23 copies

Loys Le Roi
On printing
1974, 11 pp, 120 copies

The Copper Plate Printer
1975, 9 pp, 120 copies

THE ROSEMARY PRESS
Dr Albert Outram
Formerly 75 Belvoir Drive
Aylestone Road, Leicester

Albert Outram
A Thematic Anthology of Prose and Verse
1948, 154 pp + 2 engraved portraits, approx 109 copies (some not made up)

Albert Outram
Exercises in Punctuation and Rules for Punctuation
1951, 60 copies

Albert Outram
The Limitations of Science
1960, iv; 16 pp, 64 copies

Albert Outram
Mysticism
Not completed

Albert Outram
A Short History of Philosophy
Not completed

SAINT ANTHONY PRESS
C. P. Healey
Formerly 10 Holbrook Road
Leicester

Oliver Bayldon
Morning
Illustrated by Chris Shorten
1959, 4 pp, 100 copies

Ray Seaton
Russia
Illustrated by Rigby Graham
1959, 12 pp, 120 copies

C. P. Healey
Sol-Ma-Ri
Illustrated by J. Stafford-Baker
1960, 12 pp, 90 copies

H. I. Bergstrom
My Island Star
1960, 8 pp

Federico Garcia Lorca
Llanto por Ignacio Sánchez Mejías
Translated by A. L. Lloyd
Illustrated with original woodcuts by Rigby Graham
1962, 20 pp, 100 copies

SILK PURSE PRESS
L. F. Taylor

A Useful Titbit
1968. No further details known

STEVEN PRESS
Zeta Alessandra and Gillian Greaves
c/o 48 Walton Street, Leicester

Gillian Greaves and Zeta Alessandra
One and One
1969, 7 pp, 20 copies

Desiderata
1972, 7 pp, 100 copies

THREOTEOTHA PRESS

Havelock Street
1965, 60 pp, 15 copies

A Print Book of Arches and Bridges in and around Leicester
1965, 83 pp, 15 copies

Pattern of Brickwork
1966, 89 pp, 16 copies

People
1967, 119 pp, 18 copies

Leicester
1968, 61 pp, 13 copies

Imbroglio
1968, 81 leaves, 18 copies

Rigby Graham
Some Notes on Woodcuts
1968, 55 leaves, 10 copies

Machines, a Book of Prints
1968, 61 pp, 14 copies

Polymorph 69
1969, 80 leaves, 17 copies

To have and to Hold
1970, 99 leaves, 18 copies

Polyphonic Vision
1971, 71 leaves, 17 copies

A Small Collection of English Woodcuts
1972, 58 pp, 15 copies

Minority Luggage
1973, 130 pp, 14 copies
Bound by Trevor Hickman

J. Burton (ed)
Integer
1974, approx 15 copies

TIE PRESS

Rob Armstrong
Yesterday
1969. Printed by Toni Savage. Tie Press is one of the various names used in Toni Savage and Duine Campbell's printing activities.

TRANSICAN
Boyd K. Lichfield
26 Kimberley Road, Leicester

Brian Patten
Walking Out
Illustrated by Pamlar Kindred
20 pp, 100 copies
The early poems of Brian Patten printed on a mixture of Japanese and English hand-made paper.

Boyd K. Lichfield
The Dissolving Cuckoo
120 copies

Boyd K. Lichfield
Old man in a garret
1970, 23 copies

TRIAD PUBLICATIONS
c/o Gordon Atkins
20 Woodfield Road, Oadby
Leicestershire

John Mason
Princes of Wales: their pageant, arms and traditions
1969, 32 pp, 200 copies

TWELVE BY EIGHT PRESS
John Mason
2 Ratcliffe Road, Leicester

John Mason
Twelve by Eight Paper: A Lecture to the Double Crown Club
1957

Douglas Martin
Papermaking: John Mason
1959, finished December 1958
Production by Orpheus Press

John Mason
Twelve by Eight
Illustrated by Rigby Graham. 1959
Published by Private Libraries Association

Twelve by Eight Papers
With illustrations by Ralph Dobson and Rigby Graham
1959, 100 copies
Published by Maggs

John Mason
Twelve Papers
1960, 12 pp, 100 copies

John Mason
Papermaking as an artistic craft: with a note on nylon paper
Illustrated by Rigby Graham
1959, 96 pp
Published by Faber & Faber
Republished 1963 by Twelve by Eight

In favour of bundling
Illustrated by Rigby Graham
1963, 16 pp, 150 copies (many spoilt)

J. H. Mason, RDI: a selection from the notebooks of a scholar printer made by his son, John Mason
Illustrated by Rigby Graham
1961, 43 pp, 950 copies

Oliver Bayldon
The Papermakers Craft
Illustrated by Rigby Graham
July 1964 (dated 1965), 19 pp, about 400 copies

J. G. Lubbock
Art and the Spiritual Life
Illustrated by J. G. Lubbock
1967, 22 pp + 20 plates

John Mason
More Papers Hand Made by John Mason
Various illustrators
1968 (dated 1967) 5 prelim + 35 specimen leaves, 140 copies

J. G. Lubbock
Aspects of Art and Science
1969, 34 pp + plates, 80 copies

J. G. Lubbock
Reflections from the Sea
Illustrated by J. G. Lubbock
1971, 26 pp, 85 copies

WIND TUNNEL IRREGULARS

Newarke Street
Various illustrators
1969, 37 leaves, 15 copies

Select Bibliography

John Carter
English Private Presses 1757–1961: catalogue of an Exhibition held at The Times Bookshop
April, 1961

Roderick Cave
The Private Press
London, 1971

Glasgow School of Art
The Page Right Printed: catalogue of an Exhibition
Glasgow, 1973

Rigby Graham
'The Private Press in Leicestershire', *Imbroglio* 1
1968

Juliet Standing
The Private Press Today; catalogue of the Exhibition at the 17th King's Lynn Festival
Wymondham, 1967

Private Press Books
1960-
(Where production details of most of the books mentioned on the previous pages can be found.)